Grasses, Sedges Restiads & Rushes

OF THE NATAL DRAKENSBERG

Grassland at about 2 000 m in the foreground with montane forest below; Cathedral Peak State Forest with Cathedral Peak in the background. (W. R. Bainbridge)

Ukhahlamba Series, Number 2

Grasses, Sedges Restiads & Rushes
OF THE NATAL DRAKENSBERG
SECOND EDITION

O.M. Hilliard
Illustrated by L.S. Davis

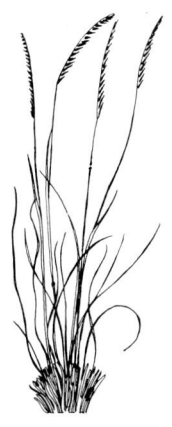

UNIVERSITY OF NATAL PRESS
PIETERMARITZBURG
1996

ISBN   0 86980 476 6 (Series)

ISBN   0 86980 535 5 (No. 2)

ISBN   0 86980 928 8 (No. 2, 2nd ed.)

Typeset by the University of Natal Press
Printed by The Natal Witness
Printing and Publishing Company (Pty) Ltd, Pietermaritzburg
Cover by On Line, Durban

CONTENTS

Plate 1a   Cobham State Forest in winter showing grassland at about 2 000m; warm aspect on the left and cool aspect on the right. Hodgson's Peak in the background. (W. R. Bainbridge)

Plate 1b   Basalt crags near Bushman's Nek gilded by an undescribed species of *Merxmuellera*. (B. L. Burtt)

PREFACE

It is my privilege to provide a few introductory comments to this second booklet in the Ukhahlamba Series.

Dr Hilliard, the author, was a particular friend of the Department of Forestry (now Environment Affairs). For many years before her retirement she was Curator of the Department of Botany Herbarium, University of Natal, Pietermaritzburg, where she gave able assistance to many field workers including our staff. A most energetic field worker herself, Dr Hilliard made numerous plant collecting forays into the State forests. Latterly she became particularly interested in the plants of the Drakensberg. For many years she has studied the high altitude plants of the southern Drakensberg and, together with B. L. Burtt, has made extensive collections of the flora of this important region, which includes parts of the only occurrence of alpine vegetation in southern Africa. This work is shortly to be published.

Dr Hilliard now lives in Scotland where she uses the facilities of the Royal Botanic Garden, Edinburgh, to continue her work on southern African plants.

Linda Davis, the illustrator, is also a former member of the Department of Botany, University of Natal.

The present booklet, which describes some of the common grasses and sedges, restiads and rushes of the Drakensberg, should be of use to anyone interested in the identification of these plants, including farmers, students and visitors to the protected areas of the Drakensberg. Here are conserved some of the most magnificent examples of the pristine grasslands that once covered much of Natal, but which are now becoming increasingly rare.

W. R. Bainbridge
Department of Environment Affairs
Pietermaritzburg

a. *Elionurus muticus*

b. *Festuca sp.*

c. *Panicum ecklonii*

d. *Alloteropsis semialata*

Plate 2   Some common Drakensberg grasses. (W.R. Bainsbridge)

INTRODUCTION

This is the second in a series of booklets intended for visitors to the Natal Drakensberg who would like to know the names of some of the plants they see, but who have no training in botany. It deals with the grasses and sedges, as well as the sedge-like restiads and rushes; these groups (families) comprise, in number of individuals though not in number of species, the principal plants of the Berg.

There is a series of keys, which will lead you step by step to the name of the plant you want to identify. The first is a key to the four families, in which 3 pairs of contrasting statements will enable you to decide whether you are dealing with a grass, a sedge, a restiad or a rush. Very few technical terms have been used, and these are explained on pages 2, 3 and 4. There are also diagrams on pages 6 and 7 to show the important parts of grass and sedge plants. Further explanatory notes appear on pages 8, 49, 66 and 67, preceding the keys to the genera.

The key to the genera of grasses is divided into 4 parts: an initial, illustrated, key will guide you to one of four groups of grasses that have certain sets of characters in common, and this key is followed by separate keys to the genera in each group. Once you have arrived at a generic name, turn to the relevant page, where you will find an illustration of a species of that genus together with descriptive notes. Usually only one species in each genus is illustrated. When two or more species of a genus occur in the Berg, their names are listed and the descriptive notes will help you to distinguish them.

There is only one key to the genera of sedges, and another to the rushes; the restiads are represented by but a single genus. As in the grasses there are illustrations and notes designed to help you determine the name of your specimen.

This booklet is a field guide, meant to be used during walks; carry it in your pocket together with a x8 or x10 hand lens. The scale lines on the drawings represent 1mm.

1

BOTANICAL TERMS

acute	sharply pointed
acuminate	gradually tapering to a long point
annual	a plant that completes its life cycle in one year. All or most of the shoots will bear flowers and no runners are present at the base of the plant
awn	slender bristle-like projection from the back or tip of the glumes and lemmas in some grasses
awned	provided with an awn, in contrast to awnless
axis	main stem of the plant, or that part of the inflorescence or spikelet that bears the flowers
bisexual	having both stamens and pistil; hermaphrodite
blade	that part of the leaf above the sheath
bract	a small leaf, situated below the inflorescence or flower, often reduced to a scale
callus	hard projection at the base of the grass floret or spikelet
culm	the grass or sedge stem
digitate	divided into parts like the fingers of a hand
entire	without toothing or division
floret	a small flower
glabrous	without hairs
glaucous	blue-green
glumaceous	resembling a glume; said of perianth parts in restiads and rushes
glume	two (or rarely one missing) empty bracts at the base of the grass spikelet; also used for the bract subtending each sedge flower
inflorescence	the flower-bearing part of the plant
internode	the part of the stem between the nodes

2

lemma	lower of two bracts enclosing the grass flower
ligule	a membrane or fringe of hairs at the inner junction of leaf blade and sheath
membranous	thin and dry
node	the joint in a stem and the point at which a leaf or bract arises
nutlet	the fruit in sedges
ovary	female part of the flower, enclosing one or more ovules
palea	upper of 2 bracts enclosing the grass flower
panicle	an inflorescence with branches from the main axis
pedicel	flower-stalk
perennial	a plant that persists for more than two years. Look for sterile shoots among the flowering ones, or runners, or dormant buds on the rootstock, or charred remains of last season's growth
perianth	the outer part of the flower, which encloses the stamens and pistil
pistil	the female organ of the flower consisting, when complete, of ovary, style and stigma
raceme	unbranched inflorescence with the stalks of the flowers or spikelets arising directly from the axis
rhizome	a rootstock or stem of root-like appearance, prostrate or underground, the apex progressively sending up stems or leaves
septate	divided into compartments
sessile	without a stalk
sheath	the lower part of the leaf, which encircles the stem or culm
spike	unbranched inflorescence bearing stalkless flowers or spikelets
spikelet	unit of the grass and sedge inflorescence
spike-like	resembling a spike; applied to dense racemes and panicles
stamen	male part of the flower, composed of a stalk (filament) and an anther inside which the pollen is produced
stolon	a sucker, runner or any basal branch that is disposed to root on the surface of the ground
style	the attenuated part of the pistil, between the ovary and the stigma; it may branch, hence style-branches, which bear the stigmas

subdigitate	divided into parts like the fingers of a hand, but the parts not all arising from precisely the same point
subtend	to hold from underneath as a bract does a flower
triad	group of 3 spikelets
umbel	a cluster of stalks springing from the same point, like the ribs of an umbrella
unisexual	of one sex; having stamens or pistil only
utricle	a sac surrounding the fruit in some sedges

KEY TO THE FAMILIES

1a Plants with tough, wiry solid stems with a
    tightly clasping sheath at each node but no
    leaf blade; male and female flowers borne
    on separate plants in small spikes solitary
    at the tips of the stems . . . . RESTIADS (*Restionaceae*)
1b Plants not as described above 2

2a Stamens 6, ovules 2 or more in each ovary
    chamber (stems solid or chambered; leaf
    sheath, when present, without a ligule) . . . RUSHES (*Juncaceae*)
2b Stamens 3, ovules solitary in the ovary chamber 3

3a Stems hollow between the nodes, cylindric
    or more rarely flattened; leaf sheath split and
    ending in a ligule . . . . GRASSES (*Gramineae* or *Poaceae*)
3b Stems solid, often 3-angled, sometimes cylindric;
    leaf sheaths either split or closed, without a
    ligule . . . . SEDGES (*Cyperaceae*)

Three families are sometimes mistaken for sedges; these are Eriocaulaceae, Typhaceae and Xyridaceae, each with one genus in the Berg: *Eriocaulon, Typha* and *Xyris*. All are herbs of marshy places. *Eriocaulon* has small (c. 3–10 mm diam), round heads of tightly massed minute flowers, either snow-white or blackish, each head at the tip of a nude stalk at the base of which is a tuft of grass-like leaves. *Typha* is the well-known bulrush, a coarse herb with broad grass-like leaves, and a thick stem terminating in a dense cylindrical brown spike; the flowers are unisexual, those in the lower half of the spike all female, in the upper half, all male; the flowers are associated with innumerable brown hairs or scales. *Xyris* has small (c. 5–15 mm long) elongated heads of brown overlapping bracts from which emerge ephemeral flowers, each with 3 delicate yellow petals; each head is borne at the tip of a nude stalk, at the base of which is a tuft of grass-like leaves.

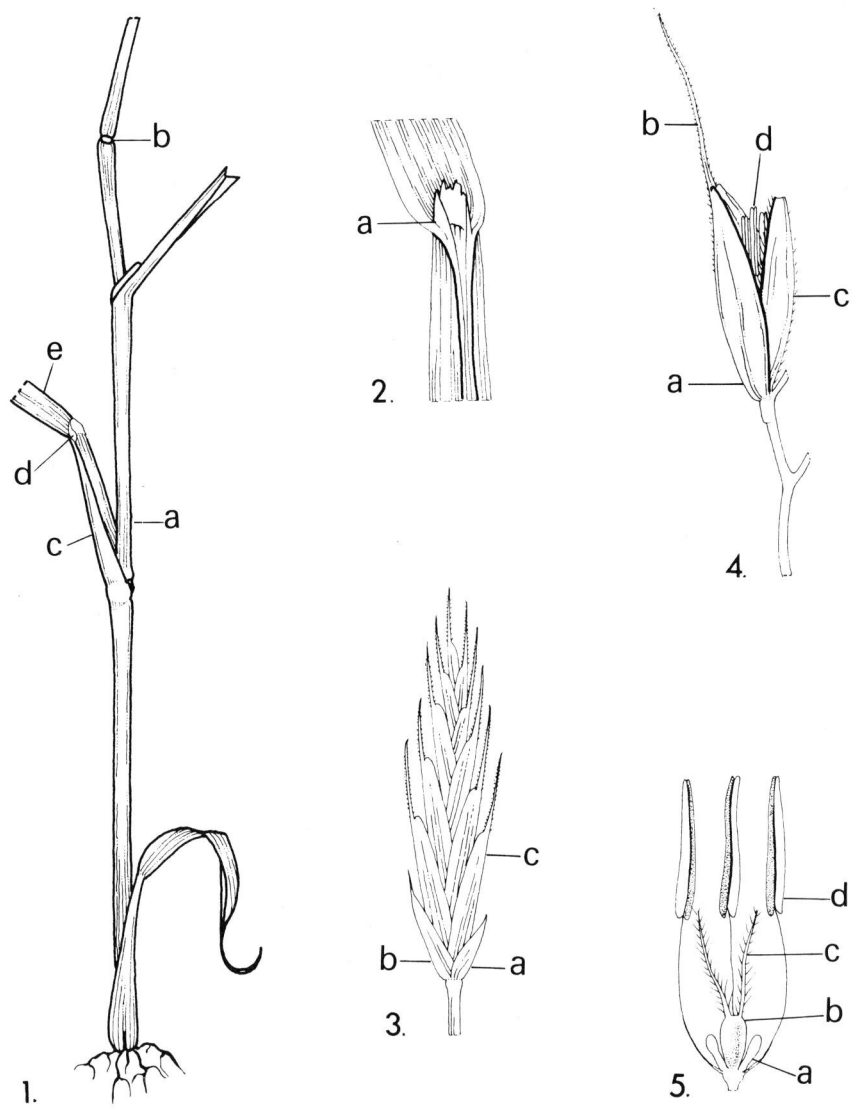

Parts of a grass plant. 1a, culm; 1b, node; 1c, leaf sheath, split to base; 1d, ligule; 1e, base of leaf blade. 2, detail of leaf, showing ligule, a. 3, a spikelet: 3a, lower glume; 3b, upper glume; 3c, lemma of the lowermost floret. 4, a floret: 4a, lemma; 4b, awn; 4c, palea; 4d, anthers. 5, floret with the lemma and palea removed: 5a, lodicule; 5b, ovary; 5c, stigma; 5d, stamen, composed of a stalk (filament) and an anther, which produces pollen (partly after Clayton, Flora of Tropical East Africa Gramineae Part I, 1970).

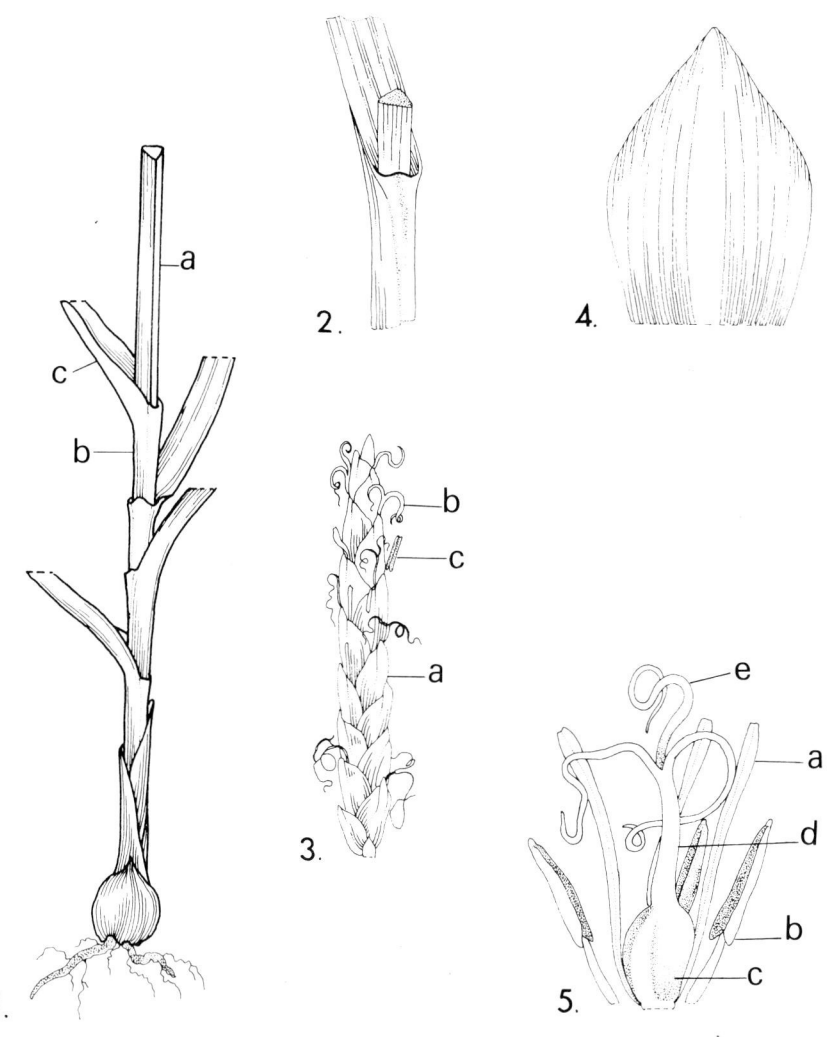

Parts of a sedge plant. 1a, 3-sided culm; 1b, tubular leaf sheath; 1c, base of leaf blade. 2, detail of culm and tubular leaf sheath. 3, a spikelet: 3a, glume; 3b, style branches (stigmas); 3c, anther. 4, a glume. 5, an hermaphrodite floret: 5a, one of the bristles or scales that constitute the perianth; 5b, stamen, composed of a stalk (filament) and anther, which produces pollen; 5c, ovary; 5d, style; 5e, style branches (stigmas).

GRAMINEAE or POACEAE

Grasses are often thought not to have flowers but of course, that is not so: they merely lack the colourful petals that one has come to associate with a flower. Grass florets (that is, small flowers) consist of two bracts, a *lemma* and a *palea*, that enclose the *stamens* (male organs) and the *pistil* (female organ); sometimes the flowers are unisexual; then only stamens or the pistil are present. Often there are two small scales at the base of the ovary; these are the *lodicules*, which are interpreted as being a highly reduced perianth or floral envelope, in other words, petals. The florets are organized into spikelets; typically a spikelet consists of 2 empty bracts or *glumes* above which are one or more florets (see diagrams on previous page). The vegetative part of the grass plant comprises the *culm* (main stem), in which are one or more joints or *nodes* (the length of culm between the nodes being the *internode*) and the leaves. In grasses, the internodes are nearly always hollow. Each leaf consists of a lower part, the leaf *sheath* (which is nearly always split to the base) and the leaf *blade*; where sheath and blade join there is often, on the upper surface, a flap of tissue or fringe of hairs; this is the *ligule*. Sometimes the ligule is a mere rim.

Use a x8 or x10 hand lens to examine grasses; you will find it rewarding to take a little trouble to learn something about them.

KEY TO GROUPS OF GRASSES

1a   Culms woody: bamboo . . . . *Thamnocalamus* p. 47
1b   Culms not woody 2

2a   Inflorescences associated with conspicuous
      and often colourful bracts (see drawings on
      p. 20 and Plate 2d) . . . . Group 1
2b   Inflorescences not associated with large bracts 3

3a   Inflorescences digitate or subdigitate . . . . Group 2

3b   Inflorescence a spike or spike-like . . . . Group 3

3c   Inflorescence an open or contracted panicle . . . . Group 4

9

GROUP 1

Inflorescences associated with conspicuous and often colourful bracts

1a Racemes solitary, shorter than and enfolded by
    the bract . . . . *Monocymbium* p.21
1b Racemes either paired or clustered, at most
    partially enfolded by the bract   2

2a Racemes distinctly paired and often standing
    well away from the bracts   3
2b Racemes in wedge-shaped clusters, each cluster
    partially enfolded by a bract . . . . *Themeda* p.23

3a Leaves aromatic (try chewing them) . . . . . . *Cymbopogon* p.20
3b Leaves not aromatic   4

4a Callus sharp (readily visible only on mature
    spikelet) . . . . *Hyparrhenia* p.20
4b Callus blunt . . . . *Andropogon* p.19

GROUP 2

Inflorescences digitate or subdigitate

1a At least some of the spikelets awned; awns
    long or short   2
1b Spikelets without awns   7

2a Racemes in pairs, borne on slender stalks clothed
    at the top in spreading hairs . . . . *Hyparrhenia* p.20
2b Racemes either not in pairs or not borne on
    slender stalks   3

3a Spikelets not paired . . . . *Alloteropsis* p.25
3b Spikelets in pairs   4

4a Callus of sessile spikelet blunt, very short   5
4b Callus of sessile spikelet sharp, at least
    2–3 mm long   6

5a Sessile spikelets with, pedicelled spikelets
    without, a bent or twisted awn . . . . *Andropogon* p.19
5b Both spikelets with a bent and twisted awn . . . . . . *Eulalia* p.19

10

GROUP 3

Inflorescence spike-like

13

GROUP 4

Inflorescence a panicle, open or contracted, but not spike-like

14

15

PHACELURUS
A genus of 9 species, Old World; 1 in s. Africa:
Phacelurus franksiae, a tufted perennial,
culms up to c. 500 mm, basal leaf sheaths
strongly compressed, blades wiry, culms and
spikelets dull red, inflorescence subdigitate,
spikelets paired, one sessile, one pedicelled.
One of the species that reddens the slopes
in November and December, after burning;
up to c. 2 000 m.

Phacelurus franksiae

ELIONURUS
Species 15, tropics and subtropics of Africa,
America and Australia; 2 in s. Africa, 1 in
the Berg: *Elionurus muticus*, a densely
tufted perennial, culms up to c. 500 mm,
leaf sheaths crimson, blades wiry, spike
c. 50 mm, spikelets paired, one sessile, one
pedicelled, dark red with silvery hairs. Damp
or marshy grassland, up to c. 2 550 m.

Elionurus muticus

17

IMPERATA

About 8 species, pantropical and warm temperate; 1 in s. Africa: *Imperata cylindrica*, a tough perennial producing tufts of stiff leaves from branching rhizomes, culms to c. 20–1 000 mm, spike thickly clad in silky white hairs. Damp places, up to c. 1 500 m.

Imperata cylindrica

MISCANTHUS

About 17 species, Old World tropics; 4 in s. Africa, 2 in the Berg. *Miscanthus sorghum* and *Miscanthus capensis* are coarse perennials, up to 2 m tall, with plume-like densely hairy inflorescences; often conspicuous on flood plains, along streambanks and in drainage lines, up to 2 500 m. In *M. capensis* the leaf blade is expanded throughout; in *M. sorghum* it is much narrowed towards the base.

Miscanthus sorghum

18

EULALIA

About 25 species, Old World tropics; 2 in
s. Africa, 1 in the Berg: *Eulalia villosa*, a
tufted perennial, culms 400–1 000 mm tall,
leaves hairy, inflorescence of 3 or more
subdigitate racemes clothed in silky hairs,
spikelets in pairs, one sessile, one pedicelled.
Grassland, up to c. 2 000 m.

Eulalia villosa

ANDROPOGON

Over 100 species, tropical and subtropical;
about 15 in s. Africa, 6 in the Berg:
*Andropogon amethystinus, A. appendiculatus,
A. distachyos, A. eucomus, A. ravus,
A. schirensis*. Tufted perennials, c. 300–
1 000 mm tall, and reaching altitudes of
c. 2 500 m; *A. amethystinus* and *A. eucomus*
occur only above 1 500 m. *A. appendiculatus*
grows in damp or marshy places and is easily
recognized by its strongly compressed glossy
yellow leaf sheaths and dull purplish subdigitate racemes. *A. distachyos* and *A. eucomus*
also favour wet places; the first has paired
racemes and broadly winged glumes, the
second digitate racemes clothed in silvery-
white hairs. *A. amethystinus, A. ravus* and
A. schirensis have paired hairy racemes; in
A. amethystinus the pedicels and internodes
of the inflorescence axis are cylindric, in
the other two they are distinctly swollen
upwards. *A. ravus* is distinctly rhizomatous
at the base, while *A. schirensis* is tufted.
All grow on rocky slopes and along streamlines.

Andropogon amethystinus

19

Cymbopogon validus

CYMBOPOGON

About 40 species, Asia and Africa; 6 in s. Africa, 1 in the Berg: *Cymbopogon validus*, a coarse tufted perennial, leaves green or glaucous, aromatic, culms 1–2 m tall, panicles of several pairs of drooping clustered racemes, each partly enclosed in a large bract. Often locally common, along valley bottoms, in drainage lines, and at margins of scrub and forest patches up to c. 2 700 m. A thatch grass.

Hyparrhenia hirta

HYPARRHENIA

Over 50 species, mostly Africa; about 20 in s. Africa, 4 in the Berg: *Hyparrhenia anamesa, H. dregeana, H. hirta, H. tamba.* Robust tufted perennials, often 1–2 m tall, never aromatic, inflorescence composed of paired racemes, each pair subtended by a long bract. The species are not easy to distinguish, and they also hybridize. Often in valley bottoms and drainage lines, sometimes in pure stands, up to c. 2 400 m. Thatch grasses.

MONOCYMBIUM

Species 3, tropical and s. Africa; 1 in the Berg: *Monocymbium ceresiiforme*, a tufted perennial, culms up to 600 mm tall, inflorescence an open panicle, the racemes solitary, each partially enclosed in a reddish-brown bract. Open slopes, up to c. 2 400 m, often locally dominant.

Monocymbium ceresiiforme

TRACHYPOGON

About 3 species, America, Africa; 1 in s. Africa: *Trachypogon spicatus*, a tufted perennial, culms reaching 1 m or more, with a characteristic ring of hairs below each node, racemes usually solitary, the spikelets in pairs, one sessile and awnless, one pedicelled and awned, the awns eventually twisting around each other. Only the fertile awned spikelets are shed at maturity; they are furnished with a sharply pointed callus that digs into clothing and flesh! Grows scattered on rocky slopes and ridges, up to 2 400 m.

Trachypogon spicatus

21

HETEROPOGON

About 6 species, tropical and subtropical; 2 in s. Africa, 1 in the Berg: *Heteropogon contortus*, a tufted perennial, culms up to 1 m tall, each terminating in a spike-like raceme, spikelets in pairs, one sessile and awned, the other pedicelled and awnless, the awns closely twisted together. Only the fertile awned spikelets are shed at maturity; as in *Trachypogon*, the sharply pointed callus digs into flesh. Mountain slopes, sometimes dominant over large areas, up to 2 600 m.

Heteropogon contortus

DIHETEROPOGON

Species 5, Africa; 2 in s. Africa, 1 in the Berg: *Diheteropogon filifolius*, a tufted perennial, leaf sheaths splitting into fibres, blades wiry, culms up to c. 600 mm tall, they and the spikelets often purplish-red, racemes usually in pairs, spikelets paired, one sessile, one pedicelled. One of the species that reddens the slopes in spring and early summer, after burning; up to c. 2 400 m.

Diheteropogon filifolius

22

THEMEDA

About 20 species, tropics and subtropics of the Old World; 1 in s. Africa; *Themeda triandra*, a tufted perennial, leaf blades red or bright brown when old, culms up to 600 mm tall, inflorescence a panicle composed of triangular units of clustered spikelets supported by conspicuous bracts flushed with red, purple and brown. Mountain slopes up to c. 2 800 m, often dominant.

Themeda triandra

DIGITARIA

About 200 species, tropical and warm temperate regions, mainly Old World; about 35 in s. Africa, 6 in the Berg: *Digitaria diagonalis, D. flaccida, D. monodactyla, D. setifolia, D. ternata, D. tricholaenoides.* *D. flaccida* has several delicate racemes arranged on a central axis, the spikelets clothed in silky purplish-red hairs; damp stony slopes, often under proteas, up to c. 2 250 m. *Digitaria monodactyla* is easily recognized by its solitary raceme with paired spikelets on a wavy axis; it favours shallow soils up to c. 2 400 m. Both are densely tufted perennials with narrow leaves 1 – 3 mm broad and culms less than 1 m tall. *D. diagonalis* is a coarse tufted perennial, culms 500 mm or more in height, usually bulbous at the base, leaves up to 10 mm broad, and an inflorescence of several long (30 – 250 mm) racemes arranged on an elongated axis; the coarse white hairs below the spikelets are characteristic; often in damp places, up to c. 1 800 m.

Digitaria flaccida

23

Digitaria monodactyla

D. setifolia is a densely tufted perennial, culms up to c. 600 mm tall, leaves wiry, racemes arranged subdigitately, spikelets clad in bright brown hairs that are diagnostic of the species; usually in damp or marshy places, up to c. 1 500 m. *D. tricholaenoides* is a perennial with a stout creeping horizontal rhizome, culms up to 600 mm tall, hairy leaves up to 8 mm wide, several racemes subdigitately arranged, spikelets clothed in long silky pink or purplish hairs, creamy when young; open grassland on thin soils, up to c. 1 700 m. *D. ternata* is a loosely tufted annual, culms mostly 200 – 500 mm long, leaves up to 8 mm broad, often spreading flat on the ground, racemes either paired or a few arranged subdigitately, spikelets densely hairy; damp rock outcrops and other damp places, often a weed along roadsides, up to c. 1 500 m.

Digitaria tricholaenoides

ALLOTEROPSIS

About 5 species, Old World; 3 in s. Africa,
1 in the Berg: *Alloteropsis semialata*, a
tufted perennial, basal leaf sheaths hairy
and often bulbous, culms mostly up to
600 mm tall, several racemes digitately
arranged, spikelets dark purple-red or dark
brown. Common, particularly in thin stony
soils, up to c. 2 000 m.

Alloteropsis semialata

BRACHIARIA

About 90 species, tropics and subtropics,
mainly Africa; about 20 in s. Africa, 2 in the
Berg: *Brachiaria marlothii* and *B. serrata*.
The first is a weed, with prostrate culms
rooting at the nodes and forming small mats,
only the flowering part turning erect; up to
c. 1 400 m. *B. serrata* is a tufted perennial,
culms up to 600 mm or more tall, the spikelets clothed in conspicuous pink, purplish
or reddish silky hairs; stony soils, up to
c. 1 800 m.

Brachiaria serrata

PASPALUM

About 250 species, tropics, mainly New World; 4 native in s. Africa, others introduced; 2 in the Berg: *Paspalum dilatatum* and *P. distichum*. *P. dilatatum* is an American species, introduced as a pasture grass and now naturalized; it is a tufted perennial with culms up to 1.8 m tall. *P. distichum* is an aggressive mat-forming perennial spreading by means of extensive underground runners, culms often very short, though sometimes reaching 250 mm. It may be indigenous. Both favour damp places, up to c. 2 000 m.

Paspalum dilatatum

OPLISMENUS

Species 5, throughout the tropics; 3 in s. Africa, 1 in the Berg: *Oplismenus hirtellus*, a perennial with rambling culms a metre or more in length, rooting and branching at the nodes, forming tangled masses in forest patches, up to c. 1 600 m.

Oplismenus hirtellus

PANICUM

About 600 species, tropical and warm temperate regions; about 40 in s. Africa, 4 in the Berg: *Panicum aequinerve, P. ecklonii, P. natalense* and *P. schinzii*. *P. aequinerve* is a delicate rambling perennial with culms up to c. 500 mm long, rooting at the nodes, the panicle very open and with relatively few spikelets; damp partially shaded places, up to c. 1 950 m. *P. ecklonii* is a tufted perennial with culms up to c. 600 mm tall, easily recognized by the toothed glumes and lower lemma; common but scattered, sometimes in marshy places, up to c. 2 450 m. *P. natalense* is also a tufted perennial, culms up to c. 600 mm tall, leaves wiry, green or glaucous; common on outcrops and thin stony soils up to c. 2 200 m. *P. schinzii* is an annual with culms up to c. 900 mm tall, leaf blades up to 20 mm broad, panicle large and open; a weed in disturbed places, up to c. 1 500 m.

Panicum ecklonii

SETARIA

About 100 species, tropics and subtropics; 14 in s. Africa, others introduced; 2 in the Berg: *Setaria obscura* and *S. pallide-fusca*. *S. obscura* is a perennial forming dense hard tufts, culms up to 1 m tall, leaves tapering to a sharp point; characterized by the paucity of bristles subtending the spikelets; damp ground along streams, up to 2 450 m. *S. pallide-fusca* is a loosely tufted annual, mostly 200–600 mm tall, the dense cylindrical inflorescence usually bright orange from the many bristles, which persist on the axis after the spikelets have fallen. This is characteristic of the genus. Damp disturbed ground, up to c. 2 000 m.

Setaria obscura

RHYNCHELYTRUM
(Sometimes included in *Melinis*)
About 15 species, mostly Africa; 1 in the
Berg: *Rhynchelytrum nerviglume*, a densely
tufted perennial, culms up to c. 600 mm
tall, leaves usually tightly rolled, glaucous,
spikelets clad in long silky hairs, creamy,
rosy or purplish. Common around rock
outcrops and in other stony ground, up to
c. 1 950 m.

Rhynchelytrum nerviglume

PENNISETUM
About 70 species, tropics and subtropics;
9 in s. Africa, others introduced, 3 in the
Berg: *Pennisetum natalense, P. sphacelatum*
and *P. thunbergii. P. natalense* is a perennial
with creeping rhizomes and culms up to
2 m tall; grows in wet places on river banks
and river islands, often partially submerged,
up to c. 1 350 m. *P. sphacelatum* and
P. thunbergii are tufted rhizomatous perennials
mostly up to c. 1 m tall, found in damp
or marshy places up to 2 900 m; in *P. sphacelatum*, the bristles on the inflorescence are
usually straw-coloured, in *P. thunbergii*
usually purple. It is characteristic of the genus
that the bristles fall with the spikelets.

Pennisetum sphacelatum

28

EHRHARTA

About 25 species, mostly SW Cape, 1 ranging to Ethiopia; 2 in the Berg: *Ehrharta erecta* and *E. longigluma*. *E. erecta* is a loosely clumped perennial, culms up to c. 1 m tall, rooting at the lower nodes, soft leaves, contracted panicle, and spikelets with glumes distinctly shorter than the lemmas; damp shady places on cliffs, under overhangs, in forest patches and *Leucosidea* scrub, up to 2 200 m. *E. longigluma* is a tufted perennial, culms up to 600 mm tall, leaves soft, panicle open, spikelets with glumes distinctly longer than the lemmas; damp gravel patches and loose gritty soil among boulders in streambeds, c. 2 200–2 900 m.

Ehrharta longigluma

PHALARIS

About 15 species, N and S temperate regions, mostly Mediterranean; 6 species in s. Africa, all introduced; 1 in the Berg: *Phalaris arundinacea*, a perennial with creeping rhizomes, stout culms up to 1,5 m tall, leaf blades up to 15 mm broad, somewhat glaucous, panicle much contracted; margins of tarns or marshy places in floodplains, up to 2 425 m. From Europe.

Phalaris arundinacea

29

ANTHOXANTHUM

About 20 species, temperate Europe, Asia, Africa; 4 native in s. Africa, 1 in the Berg: *Anthoxanthum ecklonii*, an aromatic rhizomatous perennial with culms up to c. 600 mm tall, a few tufted together, leaf blades either very narrow and then the inflorescence small with few spikelets, or the leaf blades up to 8 mm broad and the inflorescence robust; scattered on damp slopes, damp 'cliffs and around wet rock outcrops, c. 1 800 to 3 000 m.

Anthoxanthum ecklonii

ARUNDINELLA

About 45 species, tropics, mainly Asia; 1 in s. Africa: *Arundinella nepalensis*, a coarse tufted perennial with scaly rhizomes, culms up to 1,8 m tall, leaf blades stiff, panicle stiff, spikelets sometimes tinged purple, usually gaping; streambanks and wet ground below dripping cliffs, up to c. 2 000 m.

Arundinella nepalensis

TRISTACHYA

About 20 species, Africa and s. America;
7 in s. Africa, 1 in the Berg: *Tristachya
leucothrix*, a stout tufted perennial, culms
up to c. 600 mm tall, spikelets exceptionally
large, c. 25–45 mm long, golden-brown
and with very long awns, arranged in triads
on delicate drooping stalks; the spikelets
are usually clad in long hairs each arising from
a black or dark brown swollen base, but they
are sometimes few or lacking; open grassland,
sometimes locally dominant, up to c. 1 900 m.

Tristachya leucothrix

LOUDETIA

About 20 species, mainly Africa and
Madagascar; 6 in s. Africa, 1 in the Berg:
Loudetia simplex, a tufted perennial, culms
up to c. 600 mm tall, leaves narrow, glaucous,
panicle narrow, spikelets solitary or paired,
golden brown with very long awns, sometimes
glabrous, sometimes with long hairs each
arising from a black or brown swollen base;
shallow stony soils up to 2 200 m.

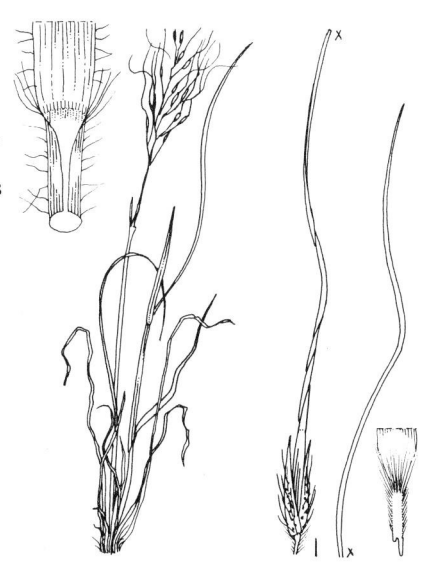

Loudetia simplex

31

AIRA
About 8 species, mainly Mediterranean;
2 in s. Africa, 1 in the Berg and there probably
native: *Aira caryophyllea*, a diminutive
annual, culms up to 50 mm tall, leaves thread-
like, panicle open, spikelets on delicate
stalks; damp soil under overhangs and at the
foot of cliffs, c. 1 980–2 750 m.

Aira caryophyllea

DESCHAMPSIA
About 60 species, temperate and cold regions
of both hemispheres, 2 in s. Africa, 1 in the
Berg: *Deschampsia caespitosa*, a densely
tufted perennial, culms up to c. 1 m tall,
leaf blades long and narrow, rigid, rough,
sharply pointed, panicle open, spikelets
purplish; marshy ground or in running water,
recorded only from the summit of the southern
Berg between c. 2 300 and 2 700 m, but also
on the high mountains of tropical Africa and
in temperate regions of both hemispheres.

Deschampsia caespitosa

32

HELICTOTRICHON

About 60 species, mainly temperate regions of northern hemisphere thence across the mountains to s. Africa; about 12 in s. Africa, 3 in the Berg: *Helictotrichon galpinii, H. natalense* and *H. turgidulum*. The species are difficult to distinguish. They are perennials forming small or large tufts, culms up to 1 m tall, and narrow panicles. All favour damp places, up to c. 2 500 m.

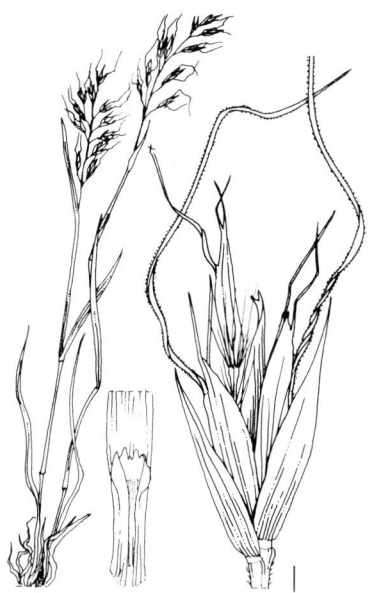

Helictotrichon natalense

MERXMUELLERA

About 16 species, s. Africa; the species are difficult to classify and some are nameless; about 9 in the Berg: *Merxmuellera aureocephala* and a nameless ally, *M. disticha, M. drakensbergensis, M. guillarmodae* and a nameless ally, *M. macowanii, M. stereophylla* and *M. stricta*. The genus is easily recognized by the tufts and fringes of hairs on the lemmas and the wiry, sharp-tipped leaves. *M. drakensbergensis* and *M. macowanii* are coarse tussock grasses with culms up to 1,5 m tall; prominent in drainage lines, along watercourses and in marshy grassland, c. 1 800– 3 000 m. *M. stereophylla* is a tufted perennial with culms up to c. 500 mm tall and dark green leaves, often conspicuous on stony ridges, rock sheets and cliffs, c. 2 000– 2 500 m. *M. aureocephala* and its unnamed ally are both tufted perennials with culms up to c. 600 mm tall, dominant on steep ridges below the summit cliffs at c. 2 150–2 500 m, the tawny-gold spikelets colouring the ridges. *M. disticha, M. stricta* and *M. guillarmodae* are also tufted perennials with culms c. 150– 450 mm tall; *M. disticha* can be recognized by its 2-ranked spikelets; gravelly patches in streambeds, crevices of rock sheets and cliffs, 2 000–3 000 m. *M. stricta* and *M. guillarmodae* both have narrow panicles and are distinguished from each other by small technical details; rock outcrops, cliffs, stony slopes, stony or sandy valley bottoms, 1 800– 3 000 m.

Merxmuellera disticha

KARROOCHLOA

Species 4, s. Africa, 1 in the Berg: *Karroochloa purpurea*, a densely tufted perennial forming small mats, leaf blades wiry, curved, beset with stiff white spreading hairs, culms 100–150 mm tall, spikelets flushed purple; on the summit plateau and near the head of Sani Pass, above 2 400 m.

Karroochloa purpurea

PENTASCHISTIS

About 70 species, mainly s. Africa; 7 in the Berg: *Pentaschistis aurea* subsp. *pilosogluma, P. exserta, P. galpinii, P. oreodoxa, P. praecox, P. setifolia, P. tysonii*. Recorded mostly above 2 000 m. They are often on moist slopes and can be subdominant at high altitudes below the summit cliffs and on the summit. All are tufted perennials with culms mostly 250–1 000 mm tall and dainty panicles; lemmas awned or awnless.

Pentaschistis angustifolia

PHRAGMITES

Species 4, cosmopolitan; 2 in s. Africa, 1 in the Berg: *Phragmites australis*, a perennial reed with creeping rhizomes, culms 1,5– 3 m tall, and plumes of silky-white spikelets; marshy ground, up to c. 1 900 m.

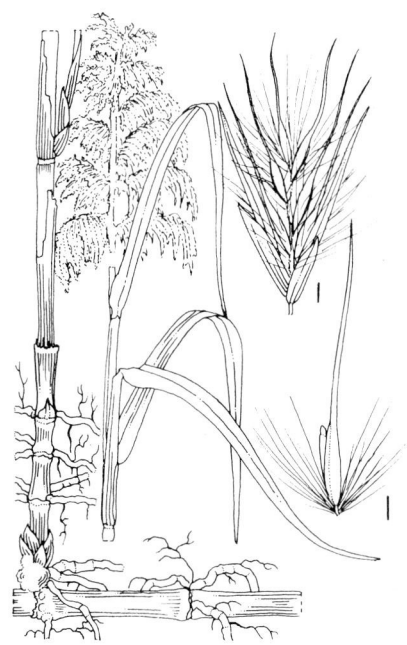

Phragmites australis

AGROSTIS

About 150–200 species, mainly temperate and cold regions in the N hemisphere, about 12 in s. Africa; at least 5 in the Berg, all characterised by the glumes gaping and persisting after the florets have fallen; all favour damp or marshy places: *Agrostis barbuligera, A. bergiana, A. continuata, A. eriantha, A. lachnantha. A. bergiana* is a delicate tufted annual with soft leaves, culms c. 15–750 mm tall, panicle pyramidal, branches hair-like, spikelets c. 2 mm long; 1 800–2 500 m. *A. lachnantha* is a loosely tufted perennial up to 600 mm tall, panicle narrow; up to c. 2 500 m. *A. continuata* is easily recognized by its spike-like panicle; up to c. 2 100 m. *A. barbuligera* and *A. eriantha* are both tufted perennials with large open panicles; in *A. barbuligera* the branches droop, in *A. eriantha* they are rigid; up to 3 000 m.

Agrostis bergiana

Aristida monticola

ARISTIDA

About 270 species, mainly tropics and subtropics; 28 in s. Africa, 2 in the Berg: *Aristida junciformis* and *A. monticola*, easily recognized by the conspicuous 3-branched awns. *A. junciformis* subsp. *galpinii* is a densely tufted perennial, culms up to c. 450 mm tall, usually with only 1 node, wiry sharp-pointed leaves, narrow panicles; stony slopes, rock sheets, rock crevices, gravelly or sandy streambeds, c. 1 800–2 700 m; *A. junciformis* subsp. *junciformis* occurs at lower altitudes, and can be distinguished by the culms with mostly 2–4 nodes. *A. monticola* is a slender straggling perennial with culms rooting from the lower nodes and forming loose tangled masses, leaves very narrow, not sharp-pointed, panicle narrow and with few spikelets; steep damp rocky slopes, streambanks and boulder beds, up to c. 2 400 m.

Stipa dregeana

STIPA ·

About 300 species, mainly temperate and warm temperate regions; 2 native in s. Africa, others introduced; 1 in the Berg: *Stipa dregeana* var. *elongata*, a perennial with a knotty rhizome, culms up to 1,5 m tall, leaf blades up to 12 mm broad, panicle large and open; lemmas hairy all over; forest floor, up to c. 1 700 m. Easily confused with *Pseudobromus silvaticus* (p. 45), which can be distinguished by its glabrous lemmas.

SPOROBOLUS

About 150 species, tropical and warm
temperate; about 40 in s. Africa, 3 in the
Berg: *Sporobolus centrifugus, S. mauritianus*
and *S. pyramidalis*, all densely tufted
perennials. In *S. centrifugus* the old leaf
sheaths are hard, glossy, yellow, the veins
not or scarcely visible, culms up to c.
800 mm tall, leaf blades wiry, spikelets 3–4 mm long,
very dark green; scattered on slopes and
ridges, sometimes in wet places, up to
c. 2 450 m. In both *S pyramidalis* and
S. mauritianus the leaf sheaths are papery, the
veins visible. *S. pyramidalis*: culms up to
1,5 m tall, panicle long and narrow, spikelets
up to 2 mm long; disturbed places, up to
1 450 m. *S. mauritianus*: culms up to 600 mm
tall, panicle long and narrow, spikelets
c. 4 mm long, very dark green; damp grass
slopes up to 1 900 m.

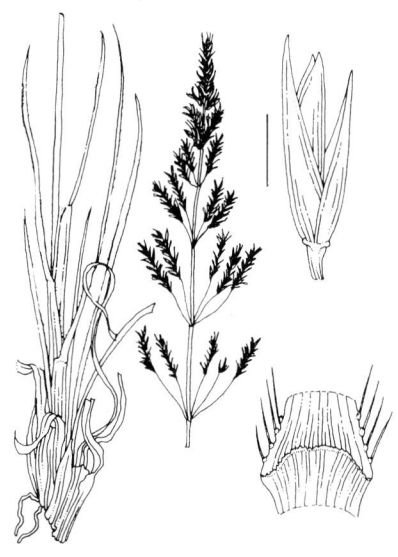

Sporobolus centrifugus

ERAGROSTIS

About 300 species, mainly tropics and subtropics; about 80 in s. Africa; 7 in the Berg:
Eragrostis aspera, E. caesia, E. capensis,
E. curvula, E. plana, E. planiculmis,
E. racemosa. E. aspera is an annual weed in
old fields, up to 1 370 m; the rest are tufted
perennials. *E. curvula* is common, especially
on damp ground, up to 2 800 m, and is
much used for reclaiming disturbed areas:
culms 300–1 200 mm tall, leaf blades very
narrow, panicle pyramidal, spikelets
4–11 x 1,5–2 mm, dark olive-grey;
E. planiculmis is difficult to distinguish from
it. *E. capensis* is easily recognized by its big
spikelets, 4–12 x 3–7 mm, straw-coloured
or purplish; thin stony soils up to 2 500 m.
E. racemosa has somewhat similar spikelets
4–10 x 2–3,5 mm, but leaden grey; thin
stony soils up to 2 500 m. *E. caesia* has
filiform leaf blades, a spike-like inflorescence,
spikelets 4–7 x 1,5–2,5 mm, very dark
grey-green; moist slopes and on the summit
plateau, 1 500–3 000 m. It may hybridize
with *E. curvula. E. plana* can be recognized
by its flattened base (often fan-shaped),
flattened culms, long narrow inflorescence,
spikelets 6–10 x 2 mm, olive-grey, the
lemmas gland-dotted; moist disturbed places,
up to 1 700 m.

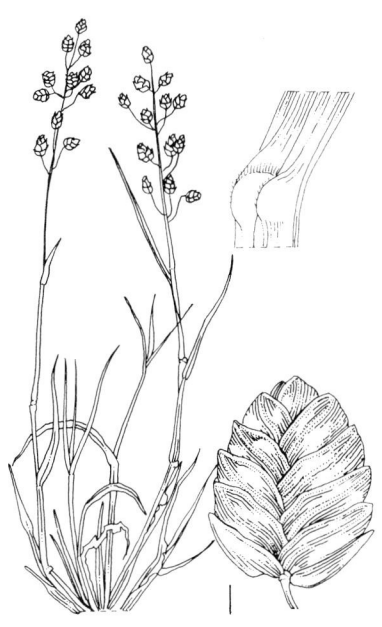

Eragrostis capensis

37

MICROCHLOA
Species 4, tropics and subtropics; 3 in s. Africa,
1 in the Berg: *Microchloa caffra*, a densely
tufted perennial, culms up to 300 mm tall,
leaf blades wiry, spike solitary, 50–150 mm
long, very slender, 1-sided, dark red-purple or
brownish-purple; rocky and partially bare
areas up to c. 2 400 m.

Microchloa caffra

RENDLIA
Species 3, tropical and south Africa, 1 in
s. Africa: *Rendlia altera*, a densely tufted
perennial, culms up to 300 mm tall, leaf
blades wiry, spike solitary, 20–50 mm long,
1-sided, dark purplish-brown or yellow;
thin stony soils up to 2 450 m, sometimes
locally dominant, and helping to colour
slopes reddish after burning.

Rendlia altera

CATALEPIS

A genus of one species, s. Africa: *Catalepis gracilis*, a tufted perennial, culms 10–300 mm tall, leaf blades wiry, inflorescence a spike-like panicle, spikelets dark greyish-purple; stony places and hard bare ground, above c. 2 300 m.

Catalepis gracilis

CYNODON

About 10 species, tropical and warm temperate; about 6 in s. Africa, some of them introduced; 2 in the Berg: *Cynodon dactylon* and *C. incompletus*. *C. dactylon* is a sward-forming perennial with both surface stolons and underground rhizomes, culms 80–400 mm tall, spikes usually 3–9, often in two whorls; the familiar Bermuda grass used for lawns, and a valuable colonizer of bare areas; found along roadsides and in weedy, trodden places, up to c. 2 400 m.
C. incompletus lacks rhizomes and the spikes are in one whorl. It favours the same habitats as *C. dactylon*.

Cynodon dactylon

HARPOCHLOA
A s. African genus of 1 species: *Harpochloa falx*, a densely tufted perennial, culms up to 600 mm tall, leaf blades expanded or folded, blunt, hard, spike solitary, 1-sided, c. 25–60 x 8 mm, curved, spikelets dark grey-green. Moist grass slopes and on the summit plateau; common.

Harpochloa falx

ELEUSINE
Species 9, mainly eastern Africa; 1 native in s. Africa, 2 introduced; 1 in the Berg: *Eleusine indica*, a robust tufted annual, culms 150–800 mm tall, inflorescence digitate or subdigitate, spikes 1-sided, spikelets green or yellowish-green; a troublesome weed, disturbed places, up to c. 1 500 m.

Eleusine indica

STYPPEIOCHLOA

A genus of 1 species, SE tropical Africa to
Natal and nearby Transkei: *Styppeiochloa
gynoglossa*, a tufted perennial forming dense
hard fibrous mats, culms c. 150–300 mm tall,
leaf blades wiry, panicle narrow, spikelets
golden-brown often heavily suffused with
purple. Damp sites over rock sheets or thin
soil, often locally dominant, 1 500– 2 700 m.

Styppeiochloa gynoglossa

FINGERHUTHIA

Species 2, s. Africa; 1 in the Berg, possibly
only south of Giant's Castle: *Fingerhuthia
sesleriiformis*, a tufted perennial, culms up
to 700 mm tall, leaf blades narrow, panicle
dense, c. 30 x 15 mm, dull purplish, bristly
from the short awns; marshy ground, 1 800–
2 800 m.

Fingerhuthia sesleriiformis

41

KOELERIA

Species 20–30, temperate regions, extending into mountainous regions of the tropics; 1 native in s. Africa: *Koeleria capensis,* a densely tufted perennial, culms up to 800 mm tall, leaf blades wiry, panicle spike-like, dense or interrupted, spikelets pale green or purplish, shining; damp slopes, up to c. 3 000 m, often locally common.

Koeleria capensis

MELICA

About 70 species, mainly temperate; 2 in s. Africa, 1 in the Berg: *Melica racemosa,* a tufted perennial, culms up to 1 m tall, inflorescence a narrow panicle or raceme, spikelets relatively few, c. 10 mm long, white faintly tinged purple, lemmas fringed with long hairs; damp and partially shaded places, up to c. 2 800 m, only very locally common.

Melica racemosa

42

STIBURUS
Species 2, s. Africa; *Stiburus alopecuroides*
and *S. conrathii*, tufted perennials, culms up
to c. 300 mm tall, leaves narrow, pointed,
panicle dense, spike-like, purplish, hairy;
damp or marshy places, sometimes over rock
sheets, often in dense stands, up to 3 000 m.
The two species are distinguished on small
technical differences.

Stiburus alopecuroides

POA
About 200 species, cosmopolitan; about 6
native in s. Africa, others introduced; at
least 3 in the Berg (the s. African species need
critical revision): *Poa annua, P. binata,
P. leptoclada. P. annua* is a tufted annual,
culms 50–150 mm tall, leaf blades soft,
bright green, tips hooded, panicle 10–100 mm
long; a cosmopolitan weed of damp and
sometimes shady places, up to 2 900 m.
P. binata is a tufted perennial, culms up to
600 mm tall, basal leaf sheaths splitting into
fibres, blades up to 5 mm broad, panicle
50–150 mm long, open; moist slopes and
other moist or marshy places, common,
sometimes mixed with *Themeda*, 1 500–
3 000 m. *P. leptoclada* is a perennial with
creeping rhizomes, culms up to 600 mm tall,
basal leaf sheaths not fibrous; panicle
contracted; wet ground or at the foot of cliffs
or under overhangs, above 2 300 m.

Poa annua

43

Festuca caprina

FESTUCA (including *Pseudobromus*).
About 400 species, temperate and subtropical
regions and mountainous areas of tropics;
about 5 in the Berg: *Festuca caprina,
F. costata, F. killickii, F. scabra* and
F. silvatica. They are all densely tufted
perennials with culms up to 1 m tall and
gaping spikelets. *F. caprina* has wiry leaves
up to 2 mm broad, and ranges in habit from
big tussocks to more or less solitary plants;
common in boulder beds, valley bottoms,
damp cliff faces and damp slopes, up to
3 000 m. The other 4 species have leaves
3–14 mm broad. In *F. scabra* the basal leaf
sheaths are broad, thickly clad in short hairs,
and the base becomes bulbous; inflorescence
a contracted panicle with all or some of the
branches bearing spikelets from the base;
slopes and ridges up to c. 2 700 m. In
F. costata and *F. killickii* the base is neither
bulbous nor hairy and the panicle is lax and
open with spikelets only on the upper half of
the branches; spikelets 12–20 mm long in
F. costata, up to 10 mm long in *F. killickii.*
F. costata favours damp slopes or marshy
places up to 2 700 m and may be locally
dominant; *F. killickii* grows along rocky
streambanks, in boulder beds and in the
crevices of damp cliffs, or, at higher
altitudes, in open grassland, 1 980–2 700 m.
F. silvatica (formerly *Pseudobromus
silvaticus*) is a tufted perennial with short
knotty rhizomes, culms up to 2 m tall, leaf
blades 6–14 mm broad, panicle large and
open, lemmas glabrous; forest floor, up
to c. 1 500 m. Bears a strong superficial
resemblance to *Stipa dregeana* (p. 36), which
is easily distinguished by its hairy lemmas.

Festuca silvatica

44

VULPIA

About 20 species, temperate and warm temperate, extending to mountainous regions in the tropics; 4 species in s. Africa, all introduced, 2 in the Berg: *Vulpia bromoides* and *V. myuros*. Both are annuals, with culms 50–600 mm tall, very narrow leaf blades and contracted panicles. In *V. bromoides*, the panicle is far exserted from the uppermost leaf sheath; in *V. myuros* it is, at most, shortly exserted. Both favour damp bare ground under overhangs or in seepage areas, 2 000–2 400 m.

Vulpia bromoides

BROMUS

About 50 species, temperate regions; 4 native in s. Africa, others introduced, 3 in the Berg: *Bromus catharticus, B. leptoclados, B. speciosus. B. cartharticus* is a tufted short-lived perennial introduced from s. America; culms up to 1 m tall, spikelets 15–40 mm long, strongly flattened, lemmas sharply keeled, awnless or with an awn up to 3 mm long; drip-line of Cave Sandstone overhangs and other damp disturbed places, up to c. 2 500 m. *B. leptoclados* and *B. speciosus* are both native species with the spikelets only weakly flattened and awns 4 mm or more long. *B. leptoclados* is a loosely tufted perennial, culms up to 1 m long, weakly erect, spikelets pale green, 5–25 (–30) mm long, lower glume 1-nerved; damp shady Cave Sandstone overhangs, c. 2 000– 2 400 m, or on the summit plateau to 3 000 m. *B. speciosus* is a tufted perennial, culms up to 1,5 m tall, spikelets marked with purple-red, 25–50 mm long, lower glume 3-nerved; moist slopes and flats, c. 1 600–2 800 m, sometimes dominant.

Bromus speciosus

45

Brachypodium bolusii

BRACHYPODIUM

About 17 species, centred on the Mediterranean, extending into temperate Europe and Asia and highlands of tropical regions; 2 native in s. Africa: *Brachypodium bolusii* and *B. flexum*. *B. bolusii* is a rhizomatous perennial producing dense tufts or mats, culms erect or straggling, mostly 300–450 mm long, leaves crowded at the base, raceme of 1–4 spikelets, these c. 20–30 mm long; under and on damp cliffs, along stream gullies, and, at high altitude, in damp loose soil in the open, 1 800–3 100 m. *B. flexum* is a weak-stemmed rhizomatous perennial forming small clumps, culms up to 1 m long, straggling, leafy for about half their length, raceme often zigzag, of 3–9 spikelets, these 15–30 mm long; damp shady places in *Leucosidea* scrub, in forest patches, and in the shelter of large boulders, up to 1 980 m.

Lolium perenne

LOLIUM

About 6 species, temperate regions of the Old World; 4 species naturalized in s. Africa, 2 in the Berg: *Lolium multiflorum* and *L. perenne*. Rye grasses, widely grown as pasture, and now found as weeds along roadsides and in other disturbed places, up to c. 3 000 m; only locally common as, for example, in Sani Pass. *L. perenne* is perennial and the lemmas are nearly always awnless; *L. multiflorum* is annual or biennial and the lemmas are usually awned.

THAMNOCALAMUS
Species 6, 5 in the Himalayas, 1 in s. Africa:
Thamnocalamus tessellatus, the Berg Bamboo,
long known as *Arundinaria tessellata*. It
produces clumps of leafy canes up to c. 2 m
tall, flowering only at long intervals; streamlines, gullies, damp rocky slopes below cliffs
and on forest margins, often locally common,
c. 1 450–2 300 m.

x 0.5

Thamnocalamus tessellatus

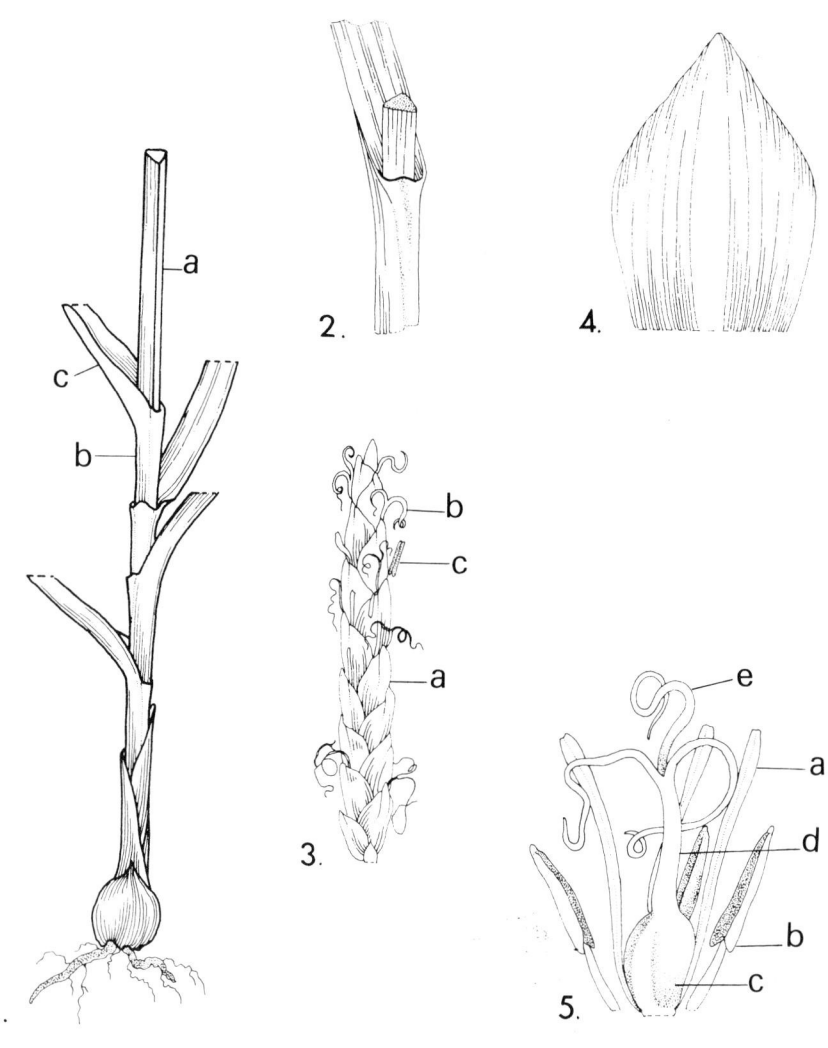

Parts of a sedge plant. 1a, 3-sided culm; 1b, tubular leaf sheath; 1c, base of leaf blade. 2, detail of culm and tubular leaf sheath. 3, a spikelet: 3a, glume; 3b, style branches (stigmas); 3c, anther. 4, a glume. 5, an hermaphrodite floret: 5a, one of the bristles or scales that constitute the perianth; 5b, stamen, composed of a stalk (filament) and anther, which produces pollen; 5c, ovary; 5d, style; 5e, style branches (stigmas).

48

CYPERACEAE

Sedges are grass-like plants, and similar terminology is used to describe the vegetative and floral parts, but grasses and sedges are easily distinguished. The sedge culm is often 3-sided; the leaf sheath is usually tubular (rarely split); each floret is either supported by a single glume or it is surrounded by a closed bract, the utricle; the perianth, if present, consists of up to 6 bristles or scales. The spikelet consists of few to many glumes, the lowermost and uppermost often empty, the others each containing an hermaphrodite or a unisexual flower. The fruit is a nutlet. Sedge genera are separated on small differences, most of which are not difficult to see with a hand lens. When you have looked at a sedge spikelet carefully and mastered the terms, the key to the genera will not be as formidable as it seems at first glance.

FIELD KEY TO THE SEDGES

ASCOLEPIS

About 20 species, most in subsaharan Africa;
2 in s. Africa, 1 in the Berg: *Ascolepis capensis*,
a tufted perennial with a very short rhizome,
culms 150–800 mm tall, scarcely 3-angled,
leaves very narrow, rolled when dry, heads
round (especially when viewed from above)
snow-white, surrounded by two or three
very long spreading bracts; marshy ground,
up to 2 250 m.

Ascolepis capensis

CARPHA

About 15 species, southern hemisphere; 5 in
s. Africa, 1 in the Berg: *Carpha filifolia*, a
densely tufted perennial, culms up to 600 mm
tall, scarcely 3-angled, leaves very narrow,
wiry, inflorescence a narrow head 15–20 mm
long supported by 1 or 2 long narrow erect
bracts overtopping it; marshes, marshy
streamsides and drainage lines, occasionally
on wet cliffs, 1 800–2 800 m, often in big
stands, the whole plant yellow-green turning
orange, very conspicuous, even at a distance.

Carpha filifolia

52

CYPERUS

About 550 species, tropical and warm
temperate; about 70 in s. Africa, 6 in the
Berg: *Cyperus marginatus, C. obtusiflorus,
C. rupestris, C. schlechteri, C. semitrifidus,
C. sphaerocephalus. C. marginatus* is easily
recognized by its stout (c. 7 mm diam)
creeping rhizomes, cylindrical leafless culms
up to c. 1 m tall and simple or compound
umbel-like inflorescence with glossy red-
brown spikelets 10–20 x 3 mm; edges of
streams and marshes, in the water, up to
c. 1 750 m. Both *C. obtusiflorus* and
C. sphaerocephalus are tufted perennials with
short thick blackish rhizomes, culms 10–
450 mm tall, tough narrow rather glaucous
leaves, and short broad spikelets in congested
heads; in *C. obtusiflorus* the spikelets are
white, in *C. sphaerocephalus*, bright yellow.
Both grow in open grassland sometimes on
stony soils, *C. obtusiflorus* up to c. 1 370 m,
C. sphaerocephalus up to 2 000 m. *C. rupestris*
forms small dense clumps, the stem bases
swollen and fibrous, culms 20–150 mm tall,
leaves wiry, spikelets 2–9, clustered in a
head, c. 6–10 mm long, up to 3 mm broad,
very dark brown or blackish; sandy or gravelly
places on and around rock sheets, or dry
stony slopes, up to 2 100 m. *C. schlechteri*
often grows with *C. rupestris* and looks rather
like it, but the plants are not tufted, there
are more spikelets in the very congested
heads, and the very dark glumes have bright
green tips; it has been recorded between
1 500 and 3 000 m. *C. semitrifidus* is not easy
to distinguish from *C. rupestris* but the
spikelets are about 4 mm broad, the styles
long, often unbranched or their branches very
short (not branches 3, longer than the style,
as in *C. rupestris*).

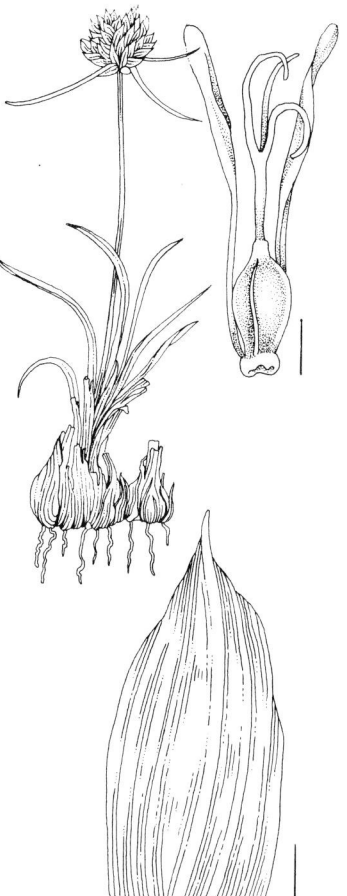

Cyperus sphaerocephalus

53

Pycreus oakfortensis

PYCREUS

About 70 species, tropical and warm temperate; about 15 in s. Africa, at least 5 in the Berg: *Pycreus flavescens, P. macranthus, P. oakfortensis, P. rehmannianus* and *P.unioloides*. *P. oakfortensis* and *P. unioloides* are both tufted perennials with very short rhizomes, culms c. 300–1 000 mm tall, long narrow leaves crowded at the base, and a terminal inflorescence of clusters of spikelets, usually 1 cluster sessile, the others short-stalked, and subtended by 2–4 long narrow bracts. *P. oakfortensis* has black spikelets; marshes up to c. 2 250 m. *P. unioloides* has golden brown spikelets; marshes up to c. 1 800m. *P. macranthus* looks not unlike *P. oakfortensis*, but it has long slender stolons; marshy places up to 1 800 m. *P. flavescens* and *P. rehmannianus* are both tufted annuals, culms up to c. 300 mm tall, long narrow leaves crowded at the base, and a terminal inflorescence of 1 sessile and several stalked clusters of spikelets. *P. flavescens* has yellow-brown spikelets sometimes tinged with red-purple; in *P. rehmannianus* they are dark brown. Both favour damp or muddy places, sometimes between boulders in streambeds or in wet depressions on rock sheets, up to 1 800 m.

MARISCUS

About 200 species, tropical and subtropical; about 30 in s. Africa, 2 in the Berg: *Mariscus congestus* and *M. drakensbergensis*. *Mariscus congestus* is a perennial with a short horizontal rhizome producing a small tuft of culms up to 1,2 m tall, leaves tough, up to 10 mm broad, inflorescence umbel-like, often large, with one sessile cluster of spikelets and 2–7 stalked clusters, the whole surrounded by broad bracts; spikelets very many, c. 8–20 x 1,5 mm, green streaked red; wet or moist hollows and damp streambanks, up to 1 980 m. *M. drakensbergensis* is a similar-looking plant but has dark brown spikelets less than 10 mm long; streambanks, c. 1 800m.

Mariscus congestus

Kyllinga pulchella

KYLLINGA

About 60 species, tropical and subtropical, mostly Africa; 14 in s. Africa, 2 in the Berg: *Kyllinga pauciflora* and *K. pulchella*. *Kyllinga pauciflora* is a delicate-looking plant with slender creeping rhizomes producing tufts of culms 80–600 mm tall, each with a few narrow (1–3 mm) leaves and a small terminal round head of up to 15 golden-green spikelets up to 6,5 mm long, the head subtended by 3 long narrow bracts; the head often produces plantlets; marshes and marshy streambanks, up to 1 900 m. *K. pulchella* has very slender stolons (less than 1 mm in diam.), culms up to c. 450 mm tall, and cylindrical dark red heads; marshy places up to 2 450 m.

55

Ficinia cinnamomea

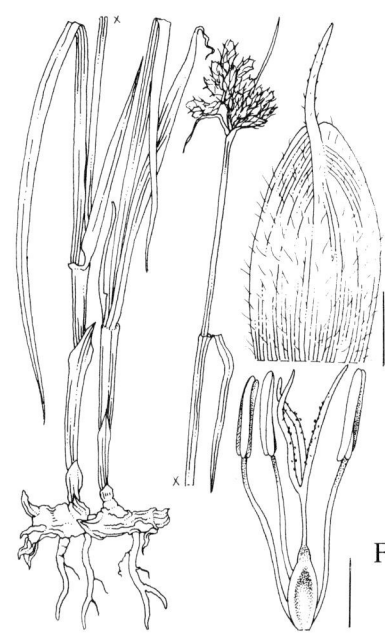

Fuirena pubescens

FICINIA

About 75 species, subsaharan Africa, mostly s. Africa, genus much in need of revision; about 5 species in the Berg: *Ficinia cinnamomea, F. filiculmea, F. nana, F. stolonifera, F. undosa*. *Ficinia cinnamomea* is a tufted perennial with dark brown leaf sheaths becoming fibrous with age, wiry culms up to 450 mm tall, wiry leaves, each culm terminating in a roundish head of chestnut-brown spikelets c. 5 mm long, overtopped by a long narrow bract; a shorter bract spreads out or down; common and forming clumps on open grass slopes, or sometimes in drainage lines, up to 2 250 m. *F. undosa* closely resembles *F. cinnamomea*, but the ripe nutlets of *F. undosa* have transverse wavy lines while those of *F. cinnamomea* are smooth.
F. stolonifera also resembles *F. cinnamomea*, but it produces slender stolons and grows singly or in small tufts; it also has fewer (up to 6) and darker spikelets, clustered together, and they are often longer, c. 6– 10 mm ; grassy slopes up to 2 100 m.
F. nana is allied to *F. stolonifera* but it is a much smaller plant with tufts of culms 10–80 mm tall, curved at the base, and spikelets only 2,5–4 mm long; damp bare ground, gravel patches and short turf, 1 980–2 500 m, southern Berg only. *F. filiculmea* has tufts of wiry culms 150–500 mm long, leaves reduced to brownish-red sheaths or occasionally produced into a wiry blade up to 30 mm long, spikelets often solitary or up to 4 clustered at the tip of the culm, 4–6 x 1,5–2 mm, reddish-brown; hanging from cliffs or growing erect on steep wet grassy banks, c. 1 900–3 000 m.

FUIRENA

About 40 species, tropical and subtropical; about 10 in s. Africa, 1 in the Berg: *Fuirena pubescens*, a stoloniferous perennial often producing dense stands, culms up to 600 mm tall, leafy, spikelets grey, hairy, spiky-looking from the projections on the glumes, in small clusters overtopped by a long bract; marshes and marshy streamsides, up to 2 300 m.

56

ELEOCHARIS

About 200 species, cosmopolitan; perhaps 8
in s. Africa, much in need of revision; 3 or 4
in the Berg: the genus easy to recognize by
its single terminal spikelet and leafless culms,
but the species difficult to identify; marshy
ground, often in shallow standing water, up to
2 450 m. The commonest species seems to be
Eleocharis dregeana or a species closely allied
to it.

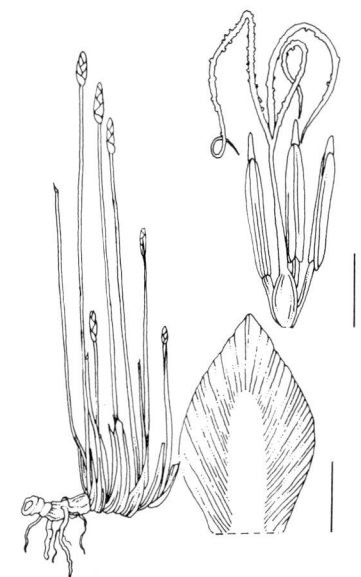

Eleocharis dregeana

SCIRPUS

This name has been used for about 40 species
in s. Africa, 10 of which occur in the Berg.
Most of these are now placed in
Schoenoplectus or *Isolepis* (below); only
2 remain, at least for the present, in *Scirpus*:
S. falsus and *S. ficinioides.*

S. falsus is a tough, tufted perennial with
short rhizomes, wiry culms 50–300 mm tall
overtopped by the wiry leaves, inflorescence
a cluster of dark spikelets overtopped by a
long culm-like bract; cliffs, rock sheets and
shallow rocky soil, 1 500–3 000 m, common.
S. ficinioides is a stouter plant than *S. falsus*,
with elongated rhizomes and culms 450–
600 mm tall; damp or marshy ground,
1 500–3 000 m.

Scirpus falsus

57

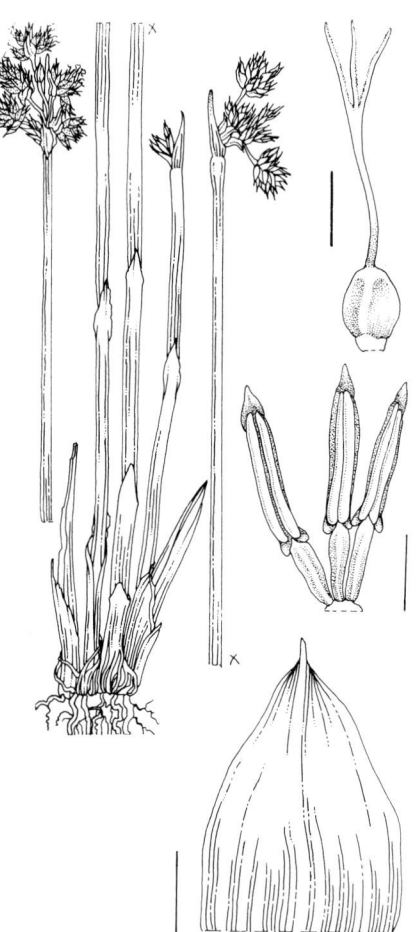

SCHOENOPLECTUS
About 50 species, temperate and tropical regions; perhaps 10 in s. Africa, 2 in the Berg: *Schoenoplectus corymbosus* [*Scirpus inclinatus*] and *Schoenoplectus decipiens*. *S. corymbosus* is a stout tough rhizomatous leafless perennial, culms up to 1 m tall, 4–8 mm in diam., green, cylindric, inflorescence subtended by a short, sharply pointed, culm-like bract and composed of clusters of reddish brown spikelets on stalks of very unequal length; marshy ground at margins of pools and streams, often partly submerged, up to c. 1 900 m. *S. decipiens* is a much more delicate-looking plant with culms up to 600 mm tall and 2–4 mm in diam., and the inflorescence overtopped by a culm-like bract c. 60–100 mm long; marshy ground, up to c. 2 350 m.

Schoenoplectus corymbosus

ISOLEPIS
About 40 species, temperate and subtropical regions, chiefly southern; about 25–30 in s. Africa, 6 in the Berg: *Isolepis angelica, I. cernua* [*Scirpus cernuus*], *I. costata* [*S. costatus*], *I. fluitans* [*S. fluitans*], *I. pellacolea, I. setacea* [*S. setaceus*]. *Isolepis fluitans* is a mat-forming perennial with creeping, branching leafy stems often rooting at the nodes, hair-like leaves, spikelets 3–6 mm long, solitary at the tips of the branchlets; marshy ground or partially submerged in pools, common, up to 3 000 m. *I. setacea* is a diminutive tufted annual, with numerous erect hair-like culms and leaves

Isolepis fluitans

58

20−60 mm long, 1−3 spikelets 2−3 mm long overtopped by a short culm-like bract; in mud, gravelly silt and moss cushions, sometimes in standing water, c. 1 800−2 600 m. *I. cernua* resembles *I. setacea*, but the surface of the nutlet is smooth (not ribbed); moist or marshy places, often near streams, up to 2 600 m. *I. angelica* is also diminutive, with culms 5−90 mm tall, a few hair-like leaves and blackish spikelets 3−7 x 2−2,5 mm, solitary or up to 3 crowded together, overtopped by the culm-like bract, but it is easily recognized by its well-developed linear rhizomes that give rise to colonies of plants in wet or moist ground above c 2 500 m. *I. costata* is a densely tufted perennial with wiry culms 10−450 mm long each with a cluster of small (up to 3 mm long) dark red-brown spikelets subtended by a very short culm-like bract, the spikelets often producing plantlets; marshy ground, up to 2 600 m. *I. pellacolea* resembles *I. costata*, but is easily recognized by its very dark brown glossy leaf sheaths, seen by parting the tufts of culms; marshy places, c. 2 000−2 400 m, and so far recorded only from the southern Berg.

RHYNCHOSPORA

About 300 species, cosmopolitan; 7 in s. Africa, 1 in the Berg: *Rhynchospora brownii,* a dainty tufted perennial, culms mostly 150−450 mm tall, slender leaves up to 3 mm broad, and narrow panicles of brown oval spikelets c. 3 mm long; damp or marshy ground around rock sheets and along streams, up to c. 2 150 m.

Rhynchospora brownii

Tetraria macowaniana

TETRARIA

About 45 species, mostly s. Africa; 2 in the
Berg: *Tetraria cuspidata* and *T. macowaniana*.
T. cuspidata is a densely tufted perennial
with short rhizomes, dark purplish-red basal
leaf sheaths, wiry culms up to c. 450 mm tall
and wiry leaves shorter than the culms,
inflorescence a very narrow panicle of few
dark reddish-brown spikelets 5–7 mm long,
c. 1 mm broad, overtopped by several long
narrow bracts; grassland, often around rock
sheets or among rocks, up to 2 600 m,
common. *T. macowaniana* is a coarser plant
with culms up to 450 mm tall, glaucous
leaves up to 6 mm broad, leaf sheaths on the
culms reddish-brown, panicle narrow, spikelets dark brown, c. 10–12 x 3 mm; tufted
on rock sheets and bare stony ground,
c. 1 800–2 200m.

FIMBRISTYLIS

About 300 species, tropical and subtropical,
especially Indomalaysia and Australia; about
15 in s. Africa, 1 in the Berg: *Fimbristylis
dichotoma*. *F. dichotoma* is a tufted perennial
with a short rhizome, culms up to 600 mm
tall, tufts of leaves up to 3 mm broad,
inflorescence umbel-like, with the spikelets
on branches of greatly varying length, spikelets 4–7 mm long, egg-shaped, bright
red-brown; marshy ground along streams,
sometimes in the water, up to 1 700 mm.

Fimbristylis dichotoma

60

BULBOSTYLIS

About 100 species, warm regions; about 16
in s. Africa, about 6 in the Berg: *Bulbostylis
densa, B. hispidula, B. humilis, B. oritrephes*
subsp. *australis, B. schoenoides, B. scleropus.*
B. *schoenoides* is a densely tufted perennial
with very short slender oblique stem bases,
becoming fibrous from the old shredded leaf
sheaths, pale rusty-brown, culms c. 70–
300 mm tall, leaves about half their length,
both hair-like, inflorescence a terminal
cluster of 1–5 brownish-black spikelets 6–
10 x 2–3 mm; damp or marshy ground, on
hillslopes, around rock sheets, in stream
gullies, 1 370–2 400 m. *B. scleropus* is
easily confused with *B. schoenoides* but it is
a coarser plant with bulbous stem bases,
broader leaf sheaths (4–5 mm, not c. 2),
blackish brown and not splitting into fibres,
leaves 1 mm broad; stony grassland, up to
c. 2 100 m. *B. oritrephes* subsp. *australis*
also resembles *B. schoenoides* but it can
be recognized at once by its conspicuous
bulbous stem bases arranged in neat
horizontal rows; stony grassland, up to
c. 2 200 m. *B. humilis* is a tufted annual 15–
100 mm tall, both culms and leaves hair-like,
leaves often overtopping the culms, culms
each bearing 1–3 spikelets at the tip, overtopped by leaf-like bracts; there are also
spikelets hidden at the base of the plant, all
pale green, sometimes flushed red; bare damp
soil, in pockets on top of rocks, between
grass tufts, and along streams, up to 2 600 m.
B. densa is also a tufted annual with hair-like
culms and leaves c. 30–150 mm or more tall,
but the spikelets are red-brown and in all but
the smallest plants they are arranged in
umbel-like inflorescences on delicate stalks
of varying length; wet sand or gravel at
edges of rock flushes or in rock hollows and
crevices, and between grass tufts in marshes,
up to c. 2 100 m. *B. hispidula* is a delicate-
looking tufted perennial, with culms 90–
300 mm tall, they and the leaves hair-like;
young plants look like annuals as they lack
the short creeping rhizome of older plants;
bare sandy places, up to 1 800 m.

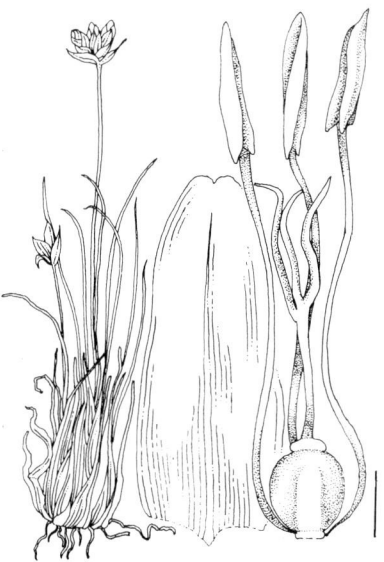

Bulbostylis schoenoides

Bulbostylis humilis

61

Scleria bulbifera

SCLERIA

About 200 species, tropical and subtropical; 23 in s. Africa, 5 in the Berg: *Scleria bulbifera, S. dieterlenii, S. dregeana, S. welwitschii, S. woodii.* The first four bear a strong superficial resemblance to one another. *S. bulbifera* has bulbous stem bases strung together horizontally, culms 300–450 mm tall, leafy, interspersed with sterile shoots composed of 3 or 4 leaves, inflorescence a spike consisting of several well-spaced clusters of bright red-brown spikelets c. 5–7 x 1–2 mm, grassland, sometimes rocky, up to c. 2 000 m. In *S. dieterlenii* the stem bases are not swollen (very slender runners are present) and the glumes are hairy; seasonally wet ground around rock sheets, c. 1 700–2 500 m. *S. welwitschii* has stout (c. 5 mm diam.) horizontal underground runners and glabrous red-brown glumes; grassland or marshy ground, 1 400–1 800 m. *S. dregeana* has thinner runners than *S. welwitschii* (c. 2–3 mm diam.) and the glumes are blackish and glabrous; marshy ground, up to 2 250 m. *S. woodii* is easily recognized by its branched inflorescence with the spikelets in small clusters on hair-like branches; grassland, up to c. 1 860 m.

SCHOENOXIPHIUM

A predominantly s. African genus; about 25 species, at least 14 in the Berg: *Schoenoxiphium bracteosum, S. buchananii, S. burttii, S. caricoides, S. distinctum, S. filiforme, S. lehmannii, S. ludwigii, S. madagascariense, S. molle, S. rufum, S. schweickerdtii, S. sparteum, S. strictum.* The genus is still being revised; many of the species are separated on critical characters beyond the scope of this booklet, but a few are relatively easy to recognize.

Schoenoxiphium schweickerdtii is a coarse perennial with a small stout rhizome, culms up to 1 m tall, a small tuft of leaves up to 10 mm broad and an erect inflorescence overtopped by the large, erect, strongly keeled primary bract; common and conspicuous, scattered on grass slopes, 1 800–2 250 m. *S. filiforme* is also common, but is inconspicuous, being a tufted perennial with very

Scleria dieterlenii

62

slender underground runners, hair-like culms up to 300 mm tall, hair-like leaves, and a small few-flowered spike-like inflorescence, the spikelets up to c. 8 x 1 mm, spreading at right angles to the axis; damp grassy cliffs and stream banks, c. 1 900 – 2 560 m. *S. bracteosum, S. caricoides, S. distinctum, S. molle, S. sparteum* and *S. strictum* are also tufted perennials of grassy places, damp cliffs (*S. distinctum, S. molle*) and damp overhangs (*S. molle*), a little more robust than *S. filiforme* and expertise is needed to identify them.

Both *S. burttii* and *S. madagascariense* are coarse tufted perennials, leaves up to 11 mm broad, culms mostly 0,6 – 1,2 m tall, the tips often arching over under the weight of the big clusters of rusty-brown spikelets borne on delicate stalks; in *S. burttii* the axes of the inflorescence are glabrous and the spikelets c. 7 x 2 mm; in *S. madagascariense* the axes are minutely hairy and spikelets c. 7 x 1 mm. Both favour damp grass slopes, grassy stream-banks, rocks at the foot of cliffs and well-lit parts of forest patches. *S. madagascariense* has been recorded only north of Giant's Castle, up to c. 1 800 m; *S. burttii* has been recorded only south of Giant's Castle, between 1 800 and 2 450 m. Both are very easily confused with *Carex zuluensis* (below), and with the common *Schoenoxiphium rufum* and *S. ludwigii*, both rather coarse tufted perennials, but with leaves often only 3 – 7 mm broad, spikelets c. 5 mm long (utricles c. 3 x 1,5 – 2 mm), in narrow pseudo-spikes, either drooping or erect, on delicate stalks; damp grassy slopes and stream gullies and forest patches, 1 800 – 2 450 m. Expertise is needed to distinguish them. *S. buchananii* is yet another tall coarse perennial, leaves up to c. 6 mm broad, the inflorescence more scanty than in the preceding 3 species, the utricle c. 5 x 1 mm and very distinctly beaked; grassy bouldery places up to c. 1 500 m.

S. lehmannii is a tufted perennial with soft leaves 2 – 4 mm broad, weakly erect culms up to c. 600 mm long, and a scanty inflorescence of small clusters or short pseudo-spikes of spikelets, often well spaced along the axis and pressed to it, or sometimes drooping on delicate stalks; forest floor, up to c. 1 800 m.

Schoenoxiphium filiforme

Schoenoxiphium schweickerdtii

63

Carex cognata

CAREX

About 1 500 – 2 000 species, cosmopolitan, especially temperate regions; about 20 in s. Africa, 8 in the Berg: *Carex acutiformis, C. austroafricana, C. bequaertii, C. cognata* var. *drakensbergensis, C. glomerabilis, C. monotropa, C. spicato-paniculata* and *C. zuluensis.*

C. acutiformis, C. austroafricana, C. bequaertii and *C. cognata* can at once be recognized by their 'cat's tails' inflorescence, the 'tails' being dense spikes drooping on long slender stalks. All are tufted perennials, often forming dense stands in wet places. *C. bequaertii* has stout culms up to 2,4 m tall, 6 – 9 spikes in each inflorescence, 100 – 150 mm long; damp places in forest, up to c. 1 500 m. The other three all have 5 – 6 spikes in each inflorescence, up to 70 mm long, and seldom reach 1 m in height; all grow in marshy places, sometimes partly submerged. In *C. acutiformis* the utricles are dull grey to brown or blackish, and strongly nerved, the beak very short (c. 0,5 mm); up to 2 350 m. In *C. cognata*, the utricle is also strongly nerved, but it is pale shining straw-colour, sometimes with rusty marks, and the beak is much longer (c. 1 – 2 mm) and sharply forked; common, c. 1 600 – 2 560 m. In *C. austroafricana* the utricle is strongly compressed, lacks conspicuous nerves and the beak is very short and blunt; also, the styles have only 2 arms (3 in the other species); up to 2 250 m.

C. glomerabilis is a rhizomatous perennial with culms up to 600 mm tall terminating in a short dense pseudo-spike, utricles shining straw-colour with rusty marks, minute projections on the shoulder and beak, style arms 2 (3 in the other species); damp or marshy ground, up to c. 2 900 m; not common. *C. monotropa* is a dwarf tufted rhizomatous perennial, the culms overtopped by the leaves, which are 40–100 mm tall; spikelets massed in dense heads, utricles yellow, strongly nerved and beaked; summit plateau only, in wet turf or wet gravel patches. *C. spicato-paniculata* and *C. zuluensis* are coarse tufted perennials, common in forest patches or in damp shady places under cliffs and big boulders, up to c. 1 800 m. They are not very easily distinguished. In *C. spicato-paniculata* the inflorescence is an open panicle, each major segment pyramidal in outline; in *C. zuluensis* each major segment is oblong in outline, the spikelets crowded together.

Carex zuluensis

Carex glomerabilis

65

RESTIONACEAE

The restiads are found chiefly in s. Africa and Australia. They much resemble some Cyperaceae in habit and inflorescence, and have creeping rhizomes and solid stems. Leaf blades are rarely present; instead at each node there is a leaf sheath split down one side and tightly rolled round the stem. The flowers are usually unisexual, and hidden by bracts; they are organized into spikes, racemes or panicles, and male and female flowers are often borne on separate plants. Each flower consists of usually 6 glumaceous perianth segments, 3 inner, 3 outer, surrounding 3 stamens in the male flower, an ovary with 1–3 styles in the female flower, where 3 rudimentary stamens may also be present. The ovary may be 1- to 3-celled, with one ovule in each chamber, and develops into a capsule or a nutlet. Only one genus is present in the Berg: *Restio*.

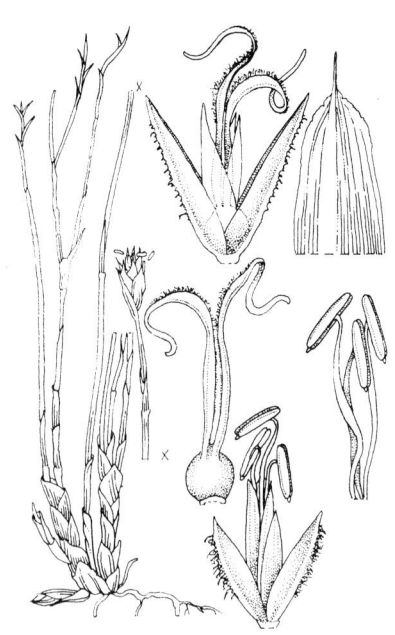

Restio schoenoides

RESTIO

About 120 species, s. Africa, mostly Cape: 4 in the Berg: *Restio galpinii, R. schoenoides* [*Ischyrolepis schoenoides*], *R. sejunctus* and an undescribed species. *R. schoenoides* is common on rock sheets and in stony ground, c. 2 100–2 500 m; it has dense tufts of wiry stems up to c. 450 mm tall, each terminating in 1–4 spikes up to c. 15 x 5 mm; the flowers are hidden by hard brown glossy overlapping bracts from which protrude either anthers or style branches, as male and female flowers are borne on separate plants. In *R. schoenoides* there are 2 style branches; in the other three species there are 3. This character will distinguish *R. galpinii* from *R. schoenoides*, which it otherwise much resembles; *R. galpinii* favours grass slopes, from c. 2 100–2 750 m, and can be locally common. The undescribed species allied to *R. galpinii* is a coarser plant that produces big clumps 1 m tall; it has been recorded in Garden Castle State Forest, at 1 950 m on a rocky streamside and at 2 300 m on an otherwise almost bare rocky basalt ridge. In *R. sejunctus* the stems are branched above the middle, the many branchlets each terminating in 1–3 spikes; marshy or damp ground at the sources of streams or in stream gullies, associated with rock sheets and boulders, up to c. 2 300 m, and probably uncommon.

JUNCACEAE

Rushes are easily confounded with sedges by the unwary, but the floral resemblance ends with the glumaceous segments that form the perianth; the rest of the flower is organized on the same plan as the lilies! There are usually six perianth segments, three outer, three inner, but the three inner are sometimes absent; they are mostly green or brown. There are usually 6 stamens and the ovary may have one chamber containing 3 ovules, or it may be divided into 3 chambers each containing many ovules; the fruit is a capsule. The stems are solid or chambered; the leaves resemble those of grasses and sedges and they may be soft and flat or hard and cylindrical; they are often reduced to an open or tubular sheath.

1a   Leaves glabrous, leaf sheath usually split down
      one side, ovary divided into 3 chambers,
      ovules many . . . . *Juncus*
1b   Leaves hairy, leaf sheath tubular, ovary
      1-chambered, ovules 3 . . . . *Luzula*

JUNCUS
About 300 species, cosmopolitan; about 25 in s. Africa, at least 9 in the Berg. *Juncus bufonius, J. dregeanus, J. effusus, J. exsertus, J. inflexus, J. mollifolius, J. oxycarpus, J. punctorius, J. tenuis.* They are all found in wet places. *J. bufonius* is a dwarf annual up to 150 mm tall, recorded from National Park at 2 100 m. *J. mollifolius* is also dwarf, but has a small, simple vertical rhizome and soft grass-like leaves; recorded from the southern Berg at c. 2 100 m, and the summit plateau. All the rest are perennials, often with well-developed rhizomes.

In *J. exsertus, J. inflexus, J. oxycarpus* and *J. punctorius*, the stems and leaves feel knotty when grasped and run through the fingers; this is because the central cavity is interrupted by transverse plates. In *J. inflexus*, the leaves are reduced to glossy brownish-black sheaths, the stems are stiffly erect, up to 1 m tall, and the cluster of flowers appears to be on one side of the stem because it is overtopped by a long sharp-pointed bract that closely resembles the stem; up to 2 600 m. *J. punctorius* has a stout stem up to 1 m tall and one leaf

Juncus exsertus

67

produced from the middle of the stem; the others are reduced to straw-coloured sheaths; the terminal inflorescence is composed of small round clusters of flowers arranged in a well-branched panicle; up to 1 800 m. Both *J. exsertus* and *J. oxycarpus* produce 3–5 leaves at the base of each stem, stems up to c. 600 mm tall, each terminating in an inflorescence composed of 1 to very many clusters of flowers; when the clusters are many, the inflorescence is well-branched. In *J. exsertus*, the capsule is longer than the glumes; in *J. oxycarpus*, it is shorter (easily seen with a hand lens); both reach 2 500 m or more.

In *J. dregeanus, J. effusus* and *J. tenuis* the stems and leaves do not have transverse joints. This will at once distinguish *J. effusus* from *J. inflexus*, which it otherwise closely resembles; up to 2 200 m. *J. dregeanus* and *J. tenuis* are both leafy. *J. dregeanus* has almost black glumes and ranges up to 2 450 m. *J. tenuis* has pale greenish glumes; it is an introduction from N America, and is an excellent colonizer of damp paths and other well-trodden places, where it forms dense mats; recorded up to 2 400 m.

LUZULA
About 80 species, cosmopolitan, especially temperate Eurasia; 1 in s. Africa: *Luzula africana*, a tufted perennial herb with stems up to c. 450 mm tall, easily recognized by its soft leaves clad in rather coarse long hairs; the flowers are crowded into a dense terminal cluster or occasionally there are several clusters borne on slender branches; damp grassland or marshy places up to 3 100 m.

Luzula africana

INDEX

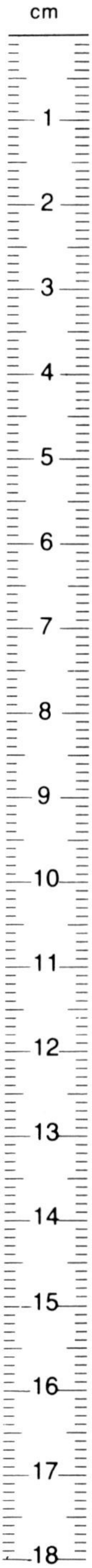

cm

1

2

3

4

5

6

7

8

9

10

11

12

13

14

15

16

17

18

UKHAHLAMBA SERIES

The Ukhahlamba Field Centre at Cathedral Peak came into being in 1977, with an agreement between the Directorate of Forestry, the University of Natal and the Natal Museum. It provides facilities for teaching and research in the Drakensberg and has contributed significantly to research development.

These booklets are being produced to help visitors to appreciate the Natal Drakensberg, and to increase public awareness of the natural resources and their management. To make them applicable over a broader area would defeat the object as they would become too large and complex; they are an aid for the development of appreciation and understanding of the Natal Drakensberg.

1. *Trees and shrubs* by O.M. Hilliard, illustrated by L.S. Davis, 2nd ed. 1992. 48pp.

2. *Grasses, sedges, restiads and rushes* by O.M. Hilliard, illustrated by L.S. Davis, 2nd ed. 1996. 80pp.

3. *Frogs and toads* by A.J.L. Lamibiris, 1988. 72pp.

4. *Flowers* by O.M. Hilliard, illustrated by L.S. Davis, 1990. 96pp.

5. *Rock paintings* by J.D. Lewis-Williams and T.A. Dowson, 1992. 68pp.